美好 編著

前言

煎炸向來是最受歡迎的烹飪方式之一，煎炸食物口感酥脆，人人喜愛，特別受年青人及小朋友歡迎。然而煎炸菜式卻予人不健康的感覺，想要享受美味的煎炸食物又怕不健康，那不如自己在家動手做！

煎炸菜式看似簡單，然而材料的搭配和火候的掌握也是一大學問。本書以煎炸為主題，介紹食材的選擇要點、基本處理方法，以及煎炸的原理和優點。書中精選三十多款家常煎炸菜式，包括炸蟹箝、香檸煎軟鴨、吉列豬扒、雜菜天婦羅、煎釀三寶等，做法簡單易學，食材容易購得，最重要是少油、少味精，不需要繁瑣工序就能在家裏做出相對健康而色香味美、甘脆可口的煎炸菜式，與家人一起分享煎炸食物的美味。

目錄

煎・蔬菜類

煎・海鮮類

煎・豬牛類

煎・雞鴨類

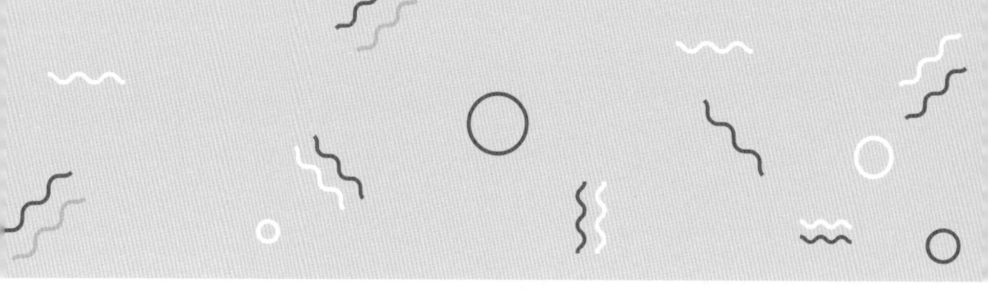

炸・蔬菜類

炸・海鮮類

炸・豬牛類

炸・雞鴨類

看圖買材料做菜

Buy ingredients according to the picture

芒果

沒有瘀痕、黑點，肉質實而軟，有濃香。

Mango

Bruise free and no brown spots, firm to touch and soft inside, pleasant fruity aroma.

鮮橙

重身，表面光滑而略黏手，橙蒂嫩綠，不易脫落。

Orange

Firm to touch and heavy for size, glossy-skinned, skin not too rough and a bit sticky to touch, fresh green stem attached to the fruit.

三色椒

表面呈光澤，硬挺，沒有破損或黑點。

Bell pepper

Glossy-skinned, firm to touch and heavy for size, cracks free and no brown spots.

檸檬

表面有油亮感覺，不焦破，帶有香味。

Lemon

Glossy-skinned, bruise free, pleasant aroma.

海苔

海藻味濃，鮮綠幼細。

Nori

Strong and rich sea weed aroma, deep green color and slender in shape.

金針菇

菇帽細小，菇柄堅挺，色澤潔白。

Enoki mushroom

Have a small cap and firm stalk with a nice cream white color.

珍珠洋葱

結實堅硬，表皮帶光澤。

Pearl onion

Firm to touch and glossy-skinned.

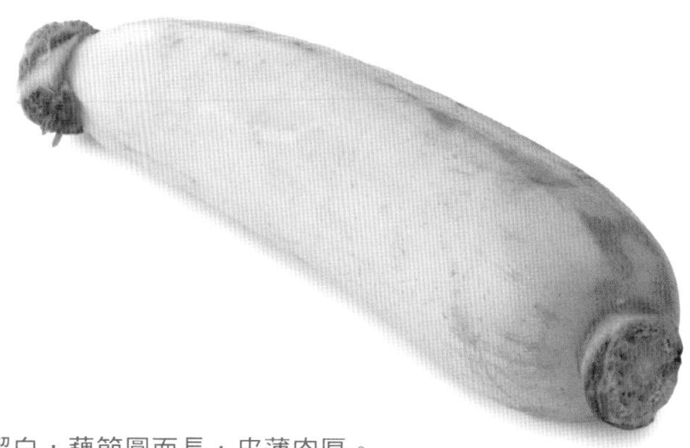

蓮藕

肉色潔白，藕節圓而長，皮薄肉厚。

The lotus root

The flesh is white, have a long, round and thick shape, thin skin but thick flesh.

急凍鴨胸

肉色粉紅，皮色潔白，沒有瘀傷。

Frozen duck breast

Pink meat and white skin, bruise mark free.

豬扒

肉質結實，有少量脂肪。

Pork chop

Firm to touch and a little fat.

牛肉粒

脂肪勻稱，肉的纖維幼細。

Beef cube

The fat marbling is evenly distributed on the meat with thin meat fibers.

牛仔骨

以三支骨為最佳部份，肉質鬆化，脂肪呈雲石般分佈。

Beef ribs

The best beef ribs are short ribs, the meat is tender with even fat marbling.

急凍蝦

蝦殼有光澤，頭身不分離。

Frozen prawn

Shiny shell, the body and the head are attached.

門鱔

肉質結實肥厚，皮光肉滑。

Pike eel

The body is firm to touch and glossy-skinned.

鱔魚

肉質結實，切口凹凸不平，魚肉有七彩光澤。

Eel

Firm to touch, the incision is rough and uneven, the flesh should look glossy.

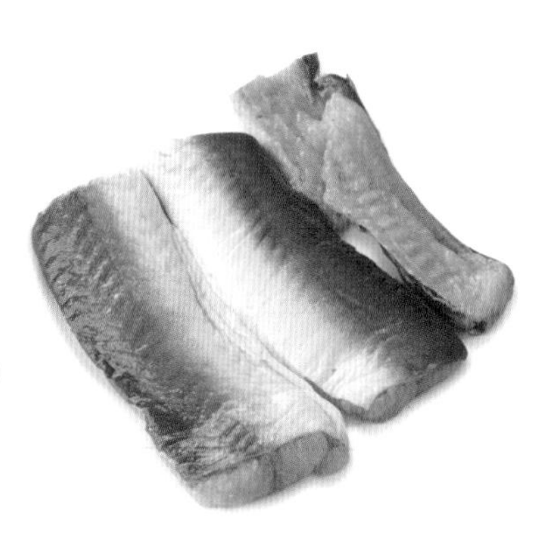

處理急凍雞翼
Tips to handle frozen chicken wings

雞翼解凍後，用少許鹽擦洗，再過清水，抹乾。
Thaw chicken wings, then rub with some salts; rinse afterwards and pat dry.

起雞翼骨（用廚剪）
Debone chicken wing with a pair of kitchen scissors

1. 用剪刀沿雞骨位下剪。
2. 小心把雞骨退出。
3. 用力扯掉雞骨。
4. 完成脫骨過程。

1. Cut along the bone of chicken wing with a pair of kitchen scissors.
2. Remove the bones carefully.
3. Pull out the bones.
4. Finish the debone process.

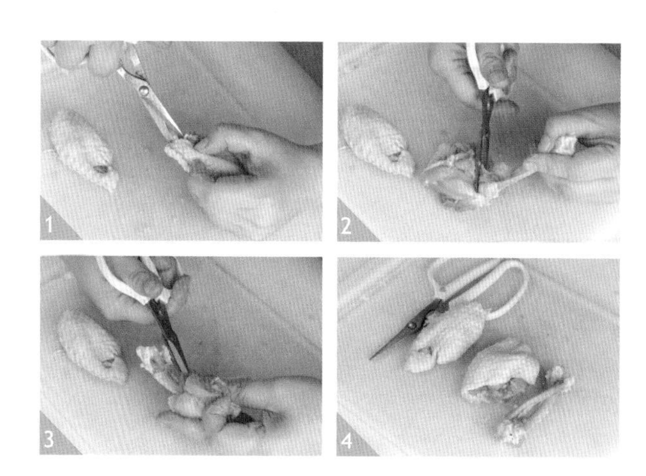

起雞翼骨（用刀）
Debone chicken wing with a knife

1. 雞翼的首末兩端用刀切開。
2. 如遇骨肉相連，可用刀把肉刮開。
3. 小心把雞骨脫出。
4. 再用刀把雞骨移去。
5. 完成脫骨過程。

1. Cut the ends of the chicken wing.
2. Separate the meat from the bone with a knife.
3. Remove bones with care.
4. Use a knife to cut the meat away from the bones.
5. Finish the debone process.

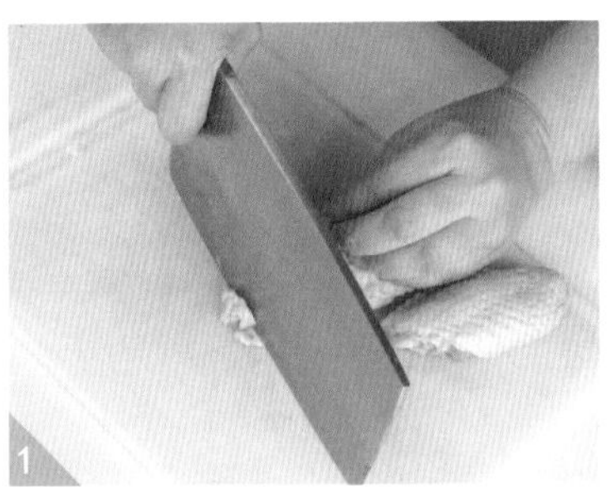

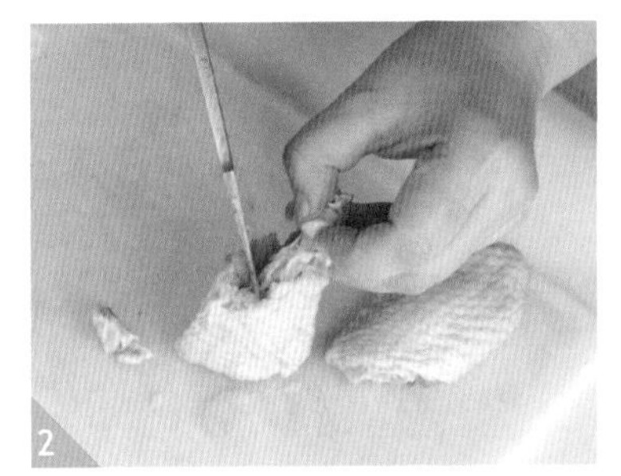

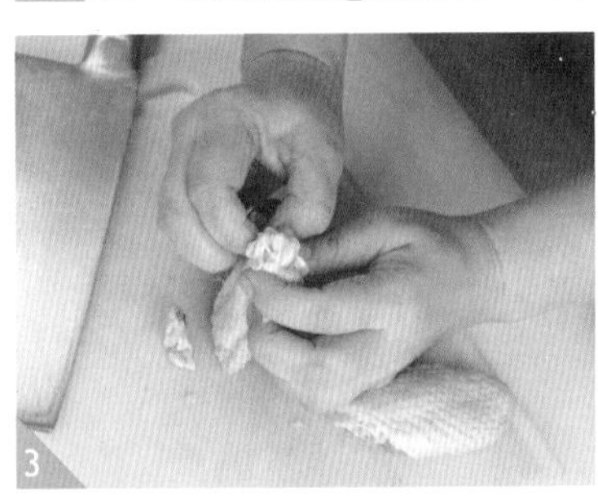

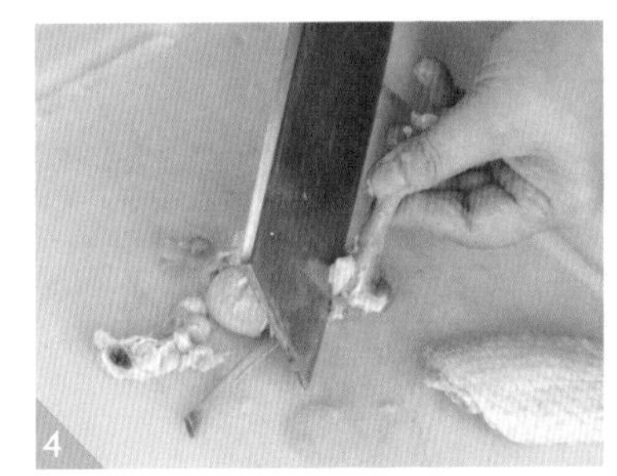

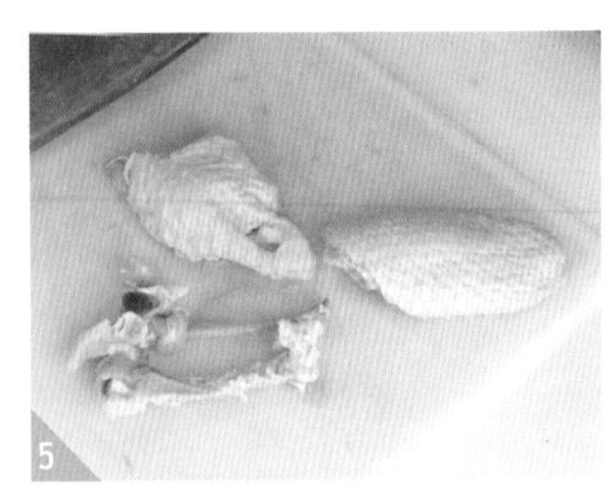

鱔魚剞花 Score the eel

鱔魚買回來，燒熱清水，熄火，把鱔魚皮向下放進熱水浸片刻，取出用小刀刮去潺液。

Boil some water and then turn off the heat. Dip the eel with the skin facing down into boiled water for a while. Use a small knife to scrap away the fluid on the eel.

1. 鱔魚洗淨後，在肉片上用刀綿密細切。
2. 在魚肉上剞直紋。
3. 然後切段。

1. Rinse eel thoroughly, then finely score the eel.
2. Score fish meat straightwise.
3. Cut eel into sections.

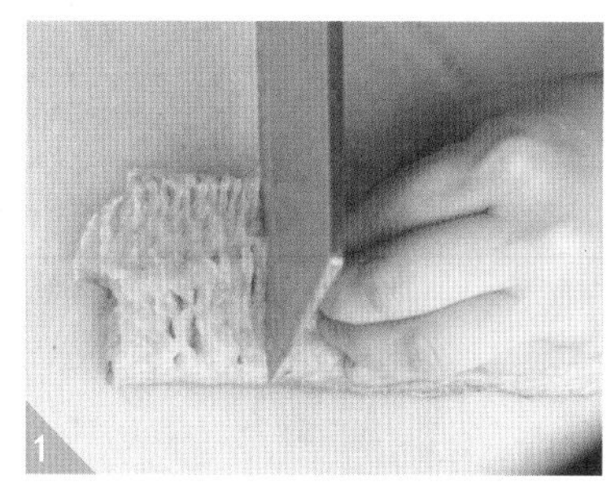

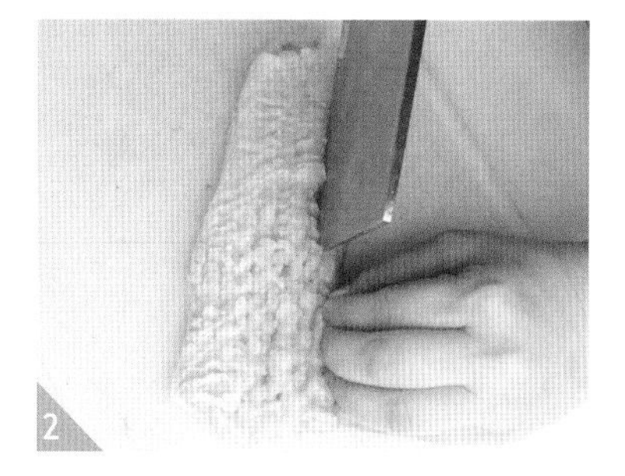

油炸 Deep-frying

即是把鑊燒熱，倒進大量生油，放入食材炸熟或上色的技法。

Deep-frying is a cooking method in which food is submerged in hot oil or fat until cooked or the food surface is colored.

提提你

- 油炸或煎的技法，往往用上大量食油，好處是外脆內軟，表面金黃。但食材表面會出現油漉漉的感覺，可放在鐵架瀝油或用廚紙輕抹油，保持乾爽質感。

煎 Pan-frying

將經糊漿處理的扁平狀原料鋪入鑊，加少量油用中小火加熱，使食材表面呈金黃色成菜的技法。

Pan-frying is a cooking method which the ingredients, coated with batter, is spread thinly in the saucepan and cooked with a little oil over low or medium heat.

Tips

- Deep-frying or pan-frying involves cooking food in a large amount of oil. This cooking method can make the food seared with a golden surface and has a characteristic crispness and soft texture inside. However, food usually becomes greasy. To keep the crunchiness, we can put the fried food on a iron rack or use kitchen papers to absorb the excess oil.

4~6 人
Serves 4~6

10 分鐘
10 minutes

煎釀三寶
Pan-fried Bell Peppers and Eggplants with Minced Dace

材料 | Ingredients

鯪魚肉茸 150 克	150g minced dace
三色甜椒各 1/2 個	1/2 pc each assorted color of bell peppers
豆腐 1 磚	1 pc beancurd
茄子 1 條	1 eggplant
豆豉 1 茶匙	1 tsp fermented black beans
蒜茸 1 茶匙	1 tsp minced garlic

入廚貼士 | Cooking Tips

- 任何蔬菜皆可。
- We can pick any kinds of vegetables in our own choice.

醃料 | Marinade

糖 1 茶匙
生粉 1 茶匙
鹽 1/2 匙
胡椒粉少許
麻油少許

1 tsp sugar
1 tsp corn flour
1/2 tsp salt
Pinch of pepper
Some sesame oil

豉汁 | Black bean sauce

糖 1 茶匙
生粉 1 茶匙
老抽 1/2 茶匙
清水 3-4 湯匙

1 tsp sugar
1 tsp corn flour
1/2 tsp dark soy sauce
3-4 tbsps water

做法 | Method

1. 甜椒洗淨切角、茄子洗淨切片、豆腐切塊，各自撲上少許生粉。
2. 鯪魚肉加入醃料拌勻待，置冰箱待 15 分鐘。放在豆腐、甜椒和茄子上輕輕壓實。
3. 熱鍋燒 3-4 湯匙油，放入蔬菜，煎至焦黃和熟透，盛起。
4. 原鑊下豆豉和蒜茸爆香，倒入豉汁煮稠，淋在三寶上。

1. Wash bell peppers, cut into wedges, wash eggplant and slice, slice beancurd, dust with corn flour.
2. Mix minced dace with marinade, marinate in the freezer for 15 minutes. Spread the mixture on beancurd, bell peppers and eggplants, tap gently.
3. Heat a wok with 3-4 tbsps of cooking oil, pan-fry vegetables and beancurd until golden brown and cooked, set aside.
4. Stir-fry fermented black beans and minced garlic in the same wok, add black bean sauce and cook till sauce is thickened, pour over vegetables and beancurd.

煎蓮藕餅

Pan-fried Minced Lotus Root Cake

材料 | Ingredients

蓮藕 200 克	200g lotus root
鯪魚肉茸 150 克	150g minced dace

醃料 | Marinade

糖 1 茶匙	1 tsp sugar
生粉 1 茶匙	1 tsp corn flour
鹽 1/2 茶匙	1/2 tsp salt
胡椒粉少許	Pinch of pepper
麻油少許	Some sesame oil

4~6 人
Serves 4~6

10 分鐘
10 minutes

芡汁 | Thickening

糖 1 茶匙	1 tsp sugar
生粉 1 茶匙	1 tsp corn flour
老抽 1/2 茶匙	1/2 tsp dark soy sauce
清水 3-4 湯匙	3-4 tbsps water

做法 | Method

1. 蓮藕洗淨，切薄片數片。其餘剁茸，撲上少許生粉。
2. 鯪魚肉加入醃料拌勻，置冰箱待 15 分鐘。放入蓮藕茸拌勻，釀夾在蓮藕片上，輕輕壓實。
3. 熱鍋燒 3-4 湯匙油，放入蓮藕片煎至焦黃和熟透，盛起。
4. 原鑊加熱，倒入芡汁煮稠，淋在蓮藕片上。

1. Wash lotus root, slice a few pieces. Finely chop the rest into puree, dust with corn flour.
2. Mix minced dace with marinade, marinate in the freezer for 15 minutes. Mix with lotus root puree, stuff and spread over lotus root slices, tap gently.
3. Heat wok with 3-4 tbsps of cooking oil, pan-fry lotus root slices until golden brown and cooked, set aside.
4. Reheat the wok, add thickening and cook till sauce is thickened, pour over lotus root slices.

入廚貼士 | Cooking Tips

- 蓮藕片容易變色，可浸在鹽水中待片刻。
- To prevent lotus roots from darkened, soak them in brine for a while.

4~6 人
Serves 4~6

10 分鐘
10 minutes

茄汁蝦碌

Pan-fried Prawns in Tomato Sauce

材料 | Ingredients

海中蝦 300 克	300g medium sized prawns
茄汁 1/3 杯	1/3 cup ketchup
蒜茸 1 湯匙	1 tbsp minced garlic
葱粒 1 湯匙	1 tbsp chopped spring onion
薑茸 1 茶匙	1 tsp minced ginger

醃料 | Marinade

生粉 1 茶匙	1 tsp corn flour
鹽 1/2 茶匙	1/2 tsp salt

調味料 | Seasonings

糖 1 湯匙	1 tbsp sugar
老抽 1 茶匙	1 tsp dark soy sauce
鹽 1/4 茶匙	1/4 tsp salt

做法 | Method

1. 海中蝦去鬚、去腳、挑去蝦腸，清洗後抹乾，並在蝦背上用刀剔開。
2. 放入鹽撈勻，決定煎蝦時才撲上生粉。
3. 熱鑊下 2-3 湯匙生油，放入海中蝦煎至八成熟兼金黃，盛起。
4. 原鑊下 1 湯匙生油，放入蒜茸和薑茸，已煎的海蝦回鑊，倒入茄汁和調味煮片刻，待收汁時，加入葱粒拌勻，上碟。

1. Cut prawns' legs and antennae, discard intestinal tracks, wash well and pat dry, cut a "Z" slit on the body.
2. Season with salt, coat with corn flour before pan-frying.
3. Heat wok with 2-3 tbsps of cooking oil, pan-fry prawns until nearly cooked and turn to golden brown color, set aside.
4. Heat the same wok with 1 tbsp of cooking oil, add minced garlic and minced ginger, add prawns, ketchup and seasonings and cook until the sauce reduced, add chopped spring onion, dish up.

入廚貼士 | Cooking Tips

- 蝦煎至八成熟而發出香味，就要盛起，否則蝦會過熟而沒有汁液，乾硬不好味。
- When the prawns are nearly cooked, we can smell the fragrant, it is time to dish up. Otherwise, over-cooked prawns are very dry and affect the taste.

雜果香芒蝦球

Pan-fried Prawns with Mango

材料 | Ingredients

海中蝦 450 克	450g medium sized prawns
芒果 1/2 個（切粒）	1/2 mango (chopped)
草莓 3-4 粒（切粒）	3-4 strawberries (chopped)
生粉 50 克	50g corn flour

醃料 | Marinade

蛋白 1/2 隻	1/2 egg white
檸檬汁 1/4 個	1/4 lemon (juiced)
生粉 1 茶匙	1 tsp corn flour
油 1 茶匙	1 tsp oil
鹽 1/2 茶匙	1/2 tsp salt
糖 1/2 茶匙	1/2 tsp sugar
胡椒粉少許	Pinch of pepper

4~6 人
Serves 4~6

10 分鐘
10 minutes

雜果汁料 | Mixed fruit juice

橙汁 1/2 杯
芒果汁 1/2 杯
檸檬汁 1 茶匙
糖 3 湯匙
吉士粉 2 茶匙

1/2 cup orange juice
1/2 cup mango juice
1 tsp lemon juice
3 tbsps sugar
2 tsps custard powder

入廚貼士 | Cooking Tips

- 雜果汁的酸甜度有異，可自行調節味道，要酸一點可加白醋或檸檬汁，想甜一些就多加點糖。
- As there are differences in the sweetness and sourness of various juice, we can adjust the taste by ourselves. We can add vinegar or lemon juice if we want to increase the sourness, or add some sugar to increase the sweetness.

做法 | Method

1. 海蝦去殼留尾，挑去蝦腸，洗淨，抹乾，並在蝦中間剔一刀，蝦尾穿過中間。
2. 把醃料拌勻，放入海蝦撈勻，醃 2-3 分鐘，取出瀝乾，沾上生粉。
3. 熱鑊下適量生油，待燒至八成滾，放入海蝦以中火半煎炸，直至蝦熟，取出瀝油。
4. 把雜果汁料放小鍋內煮滾至濃稠，熄火，再放入草莓粒和芒果粒拌勻，伴炸蝦球上桌。

1. Remove prawns' shells and keep the tails, discard intestinal tracks, wash prawns thoroughly, pat dry. Cut a "Z" slit in the middle of the body, push the tail through the slit.
2. Mix marinade and marinate prawns for 2-3 minutes, drain well, coat them with corn flour.
3. Heat a wok with cooking oil until 80% boil, pan-fry prawns over medium heat until cooked, drain well.
4. Boil mix fruit juice ingredients in a saucepan until the mixture is thickened. Turn off heat and add chopped strawberries and mango cubes, serve with prawns.

4~6 人
Serves 4~6

10 分鐘
10 minutes

煎魚餅 Pan-fried Fish Cake

材料 | Ingredients

鯪魚 1 條（450-600 克）	1 pc dace (450-600g)
臘腸 1 條	1 pc Chinese sausage
葱 1 條（切粒）	1 stalk spring onion (chopped)
冬菇 1-2 朵	1-2 pcs dried black mushrooms

醃料 | Marinade

生粉 1 湯匙	1 tbsp corn flour
鹽 1 茶匙	1 tsp salt
糖 1 茶匙	1 tsp sugar
油 1 茶匙	1 tsp oil
胡椒粉適量	Pinch of pepper

做法 | Method

1. 鯪魚去皮，洗淨，切幼粒，再剁幼。加入醃料，以順時針並大力攪至起膠。
2. 臘腸放滾水焯煮 1 分鐘，撈出，趁熱切幼粒。
3. 冬菇放入溫水浸軟，用少許生粉撈勻，放清水洗淨擠乾，切幼粒。
4. 把臘腸粒、冬菇粒、葱粒與鯪魚肉撈勻，放冰箱冷藏 10-15 分鐘。
5. 熱鑊下 2-3 湯匙生油，放入鯪魚肉按平，以中火煎至兩面金黃便可。

1. Remove skin of the dace, rinse, dice, chop, mix with marinade. Stir vigorously in clockwise direction until the mixture is thickened.
2. Poach Chinese sausage in boiling water for 1 minute, take out, finely dice sausage when it is still warm.
3. Soak dried black mushrooms in warm water until soften, mix with some corn flour. Rinse with water and pat dry, finely dice mushrooms.
4. Mix diced sausages, dried black mushrooms, spring onions and dace, put into the refrigerator for 10-15 minutes.
5. Heat a work and add 2-3 tbsps of cooking oil, put fish mixture into the wok, pat into a layer gently, pan-fry over medium heat until both sides turn into golden yellow color.

入廚貼士 | Cooking Tips

- 鯪魚肉不能沾上薑或蒜，否則魚肉會容易變霉或鬆散，不夠結實。
- The dace meat cannot smear with ginger or garlic. Otherwise, the meat will be rotten or lose its firmness.

煎封黃花

Pan-fried Yellow Croaker

材料 | Ingredients

黃花魚 1 條（約 380 克）
葱 1 條（切粒）

1 yellow croaker (about 380g)
1 sprig spring onion (chopped)

入廚貼士 | Cooking Tips

- 為免弄破魚皮，要以慢火燒熱鑊，然後加油搪勻，才把魚放進，不要移動，待其煎至乾身金黃，便會自動離鑊。
- To prevent breaking the skin of the fish, we need to heat the wok over low heat, and then add oil and spread evenly. Don't move and turn the fish over until the skin has a nice sear with a golden brown color.

4~6 人
Serves 4~6

15 分鐘
15 minutes

醃料 | Marinade

鹽 1 茶匙
生粉 1 茶匙
紹興酒 1 茶匙
胡椒粉少許

1 tsp salt
1 tsp corn flour
1 tsp Shaoxing wine
Pinch of pepper

芡汁 | Thickening

生粉 2 茶匙
糖 1 茶匙
老抽 1/2 茶匙
鹽 1/4 茶匙
清水 1/3 杯

2 tsps corn flour
1 tsp sugar
1/2 tsp dark soy sauce
1/4 tsp salt
1/3 cup water

做法 | Method

1. 魚洗淨，用布抹乾水份，加入醃料抹勻全魚，醃 5 分鐘。
2. 鑊用慢火燒熱，下 1-2 湯匙油搪勻，放入魚以中慢火煎至兩面金黃，盛起。
3. 原鑊下 1 茶匙油，放入芡汁料煮至濃稠，期間要用鑊鏟不斷打轉，避免黏底。
4. 芡汁加入葱粒稍煮滾，淋在魚上，便可。

1. Rinse fish, pat dry with a towel, marinate fish for 5 minutes.
2. Heat a wok over low heat, add 1-2 tbsps of oil, pan-fry fish over low heat until both sides turns golden brown, set aside.
3. Add 1 tsp of cooking oil into the same wok, add thickening and stir until the sauce is thickened, keep stirring the sauce to prevent it from sticking the wok.
4. Add chopped spring onion into sauce and cook for a while, pour the sauce over fish, dish up.

4~6 人
Serves 4~6

15~20 分鐘
15~20 minutes

番茄煮紅衫魚
Pan-fried Sea Bream with Tomatoes

材料 | Ingredients

紅衫魚 600 克
番茄 300 克
雞蛋 1 隻（打散）
葱 1-2 條（切段）
蒜茸 2 茶匙
鹽 1 茶匙
生粉 1-2 茶匙

600g sea bream
300g tomatoes
1 egg (whisked)
1-2 sprigs spring onion (sectioned)
2 tsps minced garlic
1 tsp salt
1-2 tsps corn flour

調味料 | Seasonings

茄汁 2 湯匙	2 tbsps ketchup
糖 3-4 湯匙	3-4 tbsps sugar
鹽 1/2 茶匙	1/2 tsp salt

做法 | Method

1. 紅衫魚劏洗乾淨，用 1 茶匙鹽擦在魚身上，醃 10-15 分鐘，撲少許生粉。
2. 鑊燒熱，下油 3-4 湯匙，放魚以中火煎至兩面金黃，盛起。
3. 番茄洗淨，切塊。
4. 把鑊燒熱，下面爆香蒜茸，加入番茄煮至汁液逼出。放入調味料煮至濃稠，加入雞蛋液和葱段兜勻，淋在魚上。

1. Gut and rinse sea bream, rub 1 tsp of salt onto the body of fish and leave for 10-15 minutes. Pat with some corn flour.
2. Heat a wok with 3-4 tbsps of oil, pan-fry sea bream over medium heat until both sides are golden, dish up.
3. Rinse tomatoes and cut into pieces.
4. Heat a wok, sauté minced garlic until fragrant, add tomatoes and cook until liquid release. Add seasoning and cook until sauce thickened. Add egg and spring onion sections and pour over the sea bream.

入廚貼士 | Cooking Tips

- 任何冰鮮魚也可以做此菜式。
- Any frozen fish could be used to make this dish.

茄汁煎豬扒

Pan-fried Pork Chops in Tomato Sauce

材料 | Ingredients

豬扒 3-6 塊	3-6 pcs pork chops
番茄 2 個	2 tomatoes
洋葱 1 個	1 onion

醃料 | Marinade

糖 1 茶匙	1 tsp sugar
生粉 1 茶匙	1 tsp corn flour
油 1 茶匙	1 tsp oil
鹽 1/2 茶匙	1/2 tsp salt
紹興酒 1/2 茶匙	1/2 tsp Shaoxing wine

4~6 人
Serves 4~6

10 分鐘
10 minutes

調味料 | Seasonings

茄汁 1/3 杯	1/3 cup ketchup
黃糖 3 湯匙	3 tbsps brown sugar
鹽 1/2 茶匙	1/2 tsp salt

做法 | Method

1. 豬扒解凍，洗淨抹乾，用刀背稍拍，加入醃料撈勻待 15 分鐘，熱鑊燒油 1-2 湯匙，放入豬扒煎熟。
2. 洋葱去衣、切絲，番茄洗淨、切角。
3. 熱鑊下 1-2 湯匙油，炒香洋葱盛起，再放進番茄煮至出汁。
4. 倒入調味和豬扒煮至汁收稠，加入洋葱絲拌勻上碟。

1. Thaw pork chops, rinse and pat dry, tap pork chops with the back of a knife, marinate for 15 minutes. Heat a wok with 1-2 tbsps of cooking oil, pan-fry pork chops till cooked.
2. Peel and shred onion, cut tomato into wedges.
3. Heat a wok with 1-2 tbsps of cooking oil, stir-fry onion and set aside. Stir-fry tomatoes until juice are released.
4. Add seasonings and pork chops, cook until sauce is thickened, toss with onion shreds, dish up.

入廚貼士 | Cooking Tips

- 用急凍的巴西豬扒，肉質腍軟，但仍要略拍，入口才鬆化。
- It is advised to use frozen pork chop from Brazil as the meat is much tender. However, be reminded to tap the pork chop to preserve the tenderness.

4~6 人
Serves 4~6

5 分鐘
5 minutes

金菇牛柳卷

Enoki Mushroom and Beef Rolls

材料 | Ingredients

肥牛肉 300 克
金菇 1 包（去根）
蒜茸 1 茶匙

300g fatty beef
1 bag Enoki mushrooms (roots removed)
1 tsp minced garlic

調味料 | Seasonings

日式燒汁 2 湯匙	2 tbsps teriyaki sauce
鹽 1/2 茶匙	1/2 tsp salt
胡椒碎 1/2 茶匙	1/2 tsp crashed black pepper
香草蒜鹽 1/2 茶匙	1/2 tsp garlic salts with herbs

做法 | Method

1. 肥牛肉從冰格轉放冰箱中解凍，用清水略沖洗，再用廚紙吸乾。
2. 金菇略沖洗，分一小撮放在肥牛肉上，捲起。
3. 熱鑊燒至冒煙，放 1-2 茶湯匙橄欖油，放入牛肉卷煎至表面轉色，封住肉汁，按需要煎 2-3 分鐘盛起。
4. 原鑊下少許油，下調味煮至收汁，淋在牛肉卷上便可。

1. Thaw fatty beef, wash, pat dry with kitchen paper.
2. Wash Enoki mushrooms, pick a pinch onto fatty beef, roll up beef.
3. Heat a wok until there is smoke, add 1-2 tsps of olive oil, pan-fry rolled beef until browned, preserve the juice inside the meat. Pan-fry for 2-3 minutes if necessary, set aside.
4. Add some cooking oil in the same wok, add seasonings and cook until sauce is thickened, pour over beef, dish up.

入廚貼士 | Cooking Tips

- 牛肉卷很快便熟，所以用易潔鑊煎既快傳熱，受熱均勻，又可省油。
- The beef rolls are cooked easily. It is better to use non-stick pan as it is easier to transmit heat and less oil is needed.

香蒜牛肉粒

Wok-fried Beef Cubes with Garlic

材料 | Ingredients

牛肉粒 300 克
蒜頭 3-4 個

300g beef cubes
3-4 cloves garlic

調味料 | Seasonings

鹽 1/2 茶匙
胡椒碎 1/2 茶匙
香草蒜鹽 1/2 茶匙

1/2 tsp salt
1/2 tsp crashed black pepper
1/2 tsp garlic salt with herbs

4~6 人
Serves 4~6

10 分鐘
10 minutes

蘸汁 | Dipping sauce

意大利陳醋適量	Some italian vinaigrette
牛扒汁或日式燒汁適量	Some steak sauce or teriyaki sauce

做法 | Method

1. 牛肉粒解凍，用清水略沖洗，再用廚紙抹乾。
2. 蒜頭去衣，切薄片，放入暖油中炸至微黃，盛起瀝油。
3. 熱鑊燒至冒煙，放 1-2 茶匙橄欖油，放入牛肉粒煎至四面轉色，肉汁封住。按需要煎 3-5 分鐘，下調味後盛起。
4. 吃前撒上香蒜片，與蘸汁伴吃。

1. Thaw beef cubes, wash, pat dry with kitchen paper.
2. Peel garlic, thinly slice, deep-fry in warm oil till slightly browned, drain well.
3. Heat a wok until there is smoke, add 1-2 tsps of olive oil, pan-fry beef cubes until browned, preserve the juice inside the meat. Pan-fry for 3-5 minutes if necessary, season and set aside.
4. Garnish with deep-fried garlic, serve with dipping sauce.

入廚貼士 | Cooking Tips

- 要封住牛肉汁，必需用高熱和短時間煎，令牛肉表面微焦，才用鑊輕壓或蓋鑊蓋弄熟。
- To preserve the juice inside the meat, brown the surface of the meat with high heat first. The action must be quick. Finish cooking by tapping the meat with a shovel or cover the meat with a lid.

4~6 人
Serves 4~6

10 分鐘
10 minutes

黑椒牛仔骨

Wok-fried Short Ribs with Black Pepper

材料 | Ingredients

牛仔骨 450 克
洋葱 1 個（切絲）
蒜茸 1 茶匙

450g short ribs
1 onion (julienne)
1 tsp minced garlic

調味料 | Seasonings

日式燒汁 2 湯匙	2 tbsps teriyaki sauce
鹽 1/2 茶匙	1/2 tsp salt
胡椒碎 1/2 茶匙	1/2 tsp crashed black pepper
香草蒜鹽 1/2 茶匙	1/2 tsp garlic salts with herbs

做法 | Method

1. 牛仔骨解凍，用清水略沖洗，再用廚紙吸乾。
2. 熱鑊燒至冒煙，放 1-2 茶湯匙橄欖油，放入牛仔骨煎至表面轉色，封住肉汁。按需要煎 5-8 分鐘盛起。
3. 原鑊下少許油，放入洋蔥絲炒香，牛仔骨回鑊，下調味煮至收汁便可。

1. Thaw short ribs, wash, pat dry with kitchen paper.
2. Heat a wok until there is smoke, add 1-2 tsps of olive oil, pan-fry short ribs until browned, preserve juice inside the meat. Pan-fry for 5-8 minutes if necessary, set aside.
3. Add some cooking oil in the same wok, stir-fry shredded onion, add short ribs, season and cook until sauce is thickened.

入廚貼士 | Cooking Tips

- 牛仔骨不要揀取過瘦或筋多的位置，太多肥膏也不好，不是太韌就是太肥了。
- Pick short ribs with care. Do not pick short ribs which are too lean, too fat or too fibrous. Otherwise, the ribs will be too greasy or unchewable.

4~6 人
Serves 4~6

10 分鐘
10 minutes

欖角豆豉雞

Pan-fried Chicken with Chinese Olives and Black Beans

材料 | Ingredients

光雞 1/2 隻	1/2 gutted chicken
辣椒 1 隻（切碎）	1 chili (chopped)
欖角 4-5 粒	4-5 cloves Chinese olives
豆豉 1 茶匙	1 tsp fermented black beans
蒜茸 1 茶匙	1 tsp minced garlic

醃料 | Marinade

糖 1 茶匙	1 tsp sugar
生粉 1 茶匙	1 tsp corn flour
油 1 茶匙	1 tsp oil
鹽 1/2 茶匙	1/2 tsp salt
紹興酒 1/2 茶匙	1/2 tsp Shaoxing wine
薑汁 1/2 茶匙	1/2 tsp ginger juice

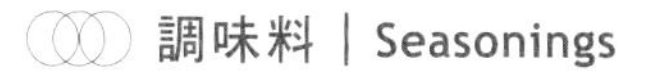

調味料 | Seasonings

生粉 1 茶匙	1 tsp corn flour
糖 1/2 茶匙	1/2 tsp sugar
鹽 1/4 茶匙	1/4 tsp salt
清水 3 湯匙	3 tbsps water

做法 | Method

1. 光雞洗淨，抹乾斬件，加入醃料拌勻待 10 分鐘。
2. 熱鑊下油，放入雞件以中火煎至八成熟，盛起。
3. 熱鑊下 2 茶匙油，放入欖角、豆豉、辣椒碎和蒜茸爆香，放入雞塊爆香，灒酒。倒入調味拌勻至濃稠收汁，上碟。

1. Rinse chicken, pat dry and cut into pieces, marinate for 10 minutes.
2. Heat a wok with cooking oil, pan-fry chicken over medium heat until nearly cooked, dish up.
3. Heat wok with 2 tsps of cooking oil, stir-fry with Chinese olives, black beans, chili and minced garlic. Add chicken pieces, sizzle some wine over the wok. Add seasonings and cook until sauce is thickened, dish up.

入廚貼士 | Cooking Tips

- 雞件要保留雞皮，肉質才幼滑，但雞油比較多且肥膩，所以用易潔鑊處理比較好。
- Keep the skin with chicken pieces will ensure that the meat is tender. As chicken fat is very greasy and hard to get rid of, it is advised to cook with a non-stick pan.

洋葱煎雞扒

Pan-fried Chicken with Onion

材料 | Ingredients

急凍雞扒（無骨雞腿肉）1 塊（約 200 克）	1 frozen chicken fillet (boneless chicken thigh) (about 200g)
紫洋葱 1 個	1 purple onion
乾葱 4-5 粒	4-5 cloves shallot
蒜頭 1 粒	1 clove garlic

醃料 | Marinade

蛋白 1 隻	1 egg white
糖 1 茶匙	1 tsp sugar
生粉 1 茶匙	1 tsp corn flour
油 1 茶匙	1 tsp oil
鹽 1/2 茶匙	1/2 tsp salt
紹興酒 1/2 茶匙	1/2 tsp Shaoxing wine
胡椒粉少許	Pinch of pepper
麻油少許	Some sesame oil

4~6 人
Serves 4~6

20 分鐘
20 minutes

調味料 | Seasonings

燒汁 1 湯匙	1 tbsp teriyaki sauce
味醂 1 湯匙	1 tbsp mirin
蒜鹽 1/2 茶匙	1/2 tsp garlic salt
清水 2-3 湯匙	2-3 tbsps water

做法 | Method

1. 雞扒解凍，洗淨抹乾，在厚肉部份用刀剁數下，加入醃料撈勻待10分鐘。
2. 紫洋葱、乾葱和蒜頭去衣，洗淨抹乾，按需要切絲和切塊。
3. 熱鑊下 2-3 湯匙生油，放入雞扒以中火煎至熟，並呈金黃，取出盛起。
4. 再次熱鑊下 2 茶匙生油，放入洋葱、乾葱和蒜頭煎至微焦，倒入燒汁和雞扒煮至汁乾，盛起。

1. Thaw chicken, rinse and pat dry. Cut a few slits on the thicker part of the meat, marinate for 10 minutes.
2. Peel onion, shallot and garlic, rinse and wipe dry, shred or cut into slices or shreds.
3. Heat a wok with 2-3 tbsps of cooking oil, pan-fry chicken over medium heat until cooked and turn to golden brown, set aside.
4. Heat wok with 2 tsps of cooking oil, brown onion, shallot and garlic, add teriyaki sauce and chicken and cook until sauce is thickened, dish up.

入廚貼士 | Cooking Tips

- 煎雞扒時可多放點油，火力用兩段，先以小火煎至差不多熟，再轉中火取色。
- Pan-fry chicken with a bit more oil. At the beginning, pan-fry over low heat. When chicken is almost cooked, turn to use medium heat to brown it.

4~6 人
Serves 4~6

10 分鐘
10 minutes

煎釀雞翼

Pan-fried Stuffed Chicken Wings

材料 | Ingredients

雞翼 10-12 隻
泰國蘆筍 10-12 條
蟹柳 5-6 條
韭菜 5-6 條
菠蘿 2 片

10-12 chicken wings
10-12 Thailand asparagus
5-6 crab sticks
5-6 stalks leek
2 slices pineapple

醃料 | Marinade

蛋白 1 隻	1 egg white
糖 1 茶匙	1 tsp sugar
生粉 1 茶匙	1 tsp corn flour
油 1 茶匙	1 tsp oil
鹽 1/2 茶匙	1/2 tsp salt
紹興酒 1/2 茶匙	1/2 tsp Shaoxing wine
胡椒粉少許	Pinch of pepper
麻油少許	Some sesame oil

做法 | Method

1. 雞翼解凍，用鹽擦後清洗，抹乾，起骨。加入醃料撈勻待 10 分鐘，瀝乾。
2. 釀入 1/2 條蟹柳、1 條蘆筍、1 條菠蘿條，再用已熟的韭菜紮實，並用牙籤定位。
3. 熱鑊燒 3-4 湯匙油，放進雞翼，以中火煎熟和呈金黃色，撈出瀝油。

1. Thaw chicken wings, rub chicken wings with salt, wash and pat dry, debone. Marinate for 10 minutes, drain well.
2. Stuff with 1/2 crab stick, 1 asparagus, 1 pineapple stick, tight chicken wings with cooked leek, fix with a toothpick.
3. Heat a wok with 3-4 tbsps of cooking oil, pan-fry chicken wings over medium heat until golden brown and cooked, drain well.

入廚貼士 | Cooking Tips

- 韭菜作裝飾，可以不用。
- The leek is for garnish only, so it is optional.

柚子蜜糖煎雞翼

Pan-fried Chicken Wings with Citron Honey

材料 | Ingredients

雞翼 10-12 隻	10-12 chicken wings
生粉適量（撲面用）	Some corn flour (for dusting)

醃料 | Marinade

蛋白 1 隻	1 egg white
糖 1 茶匙	1 tsp sugar
生粉 1 茶匙	1 tsp corn flour
油 1 茶匙	1 tsp oil
鹽 1/2 茶匙	1/2 tsp salt
紹興酒 1/2 茶匙	1/2 tsp Shaoxing wine
胡椒粉少許	Pinch of pepper
麻油少許	Some sesame oil

4~6 人
Serves 4~6

10 分鐘
10 minutes

柚子蜜汁｜Citron honey sauce

柚子蜜 2 茶匙
蒜鹽 1 茶匙
檸檬汁 1/2 個
清水 3 湯匙

2 tsps citron honey
1 tsp garlic salt
1/2 pc lemon (juiced)
3 tbsps water

做法｜Method

1. 雞翼解凍，用鹽擦後清洗，抹乾。加入醃料撈勻待 10 分鐘，瀝乾，撲上生粉。
2. 熱鑊燒 3-4 湯匙油，放進雞翼以中火煎熟，呈金黃色，取出。
3. 柚子蜜汁拌勻，放入已預熱的鑊內煮滾，倒進雞翼兜炒，上碟。

1. Thaw chicken wings, rub chicken wings with salt, wash and pat dry. Marinate for 10 minutes, drain well, dust with corn flour.
2. Heat a wok with 3-4 tbsps of cooking oil, pan-fry chicken wings over medium heat until golden brown, set aside.
3. Mix citron honey sauce ingredients, boil in a pre-heated wok, toss with chicken wings, dish up.

入廚貼士 | Cooking Tips

- 柚子蜜可改為薑蜜，別有風味。
- To make a different taste, ginger honey can be substituted for citron honey.

4~6 人
Serves 4~6

30 分鐘
30 minutes

橙汁煎鴨胸

Pan-fried Duck Breast with Orange Sauce

材料 | Ingredients

解凍鴨胸 1 塊（約 300 克）
鮮橙 1 個（裝飾用）
生粉適量（撲粉用）

1 pc duck breast (about 300g)
1 orange (for garnish)
Some corn flour (for dusting)

醃料 | Marinade

雞蛋 1 隻	1 egg
糖 1 茶匙	1 tsp sugar
生粉 1 茶匙	1 tsp corn flour
紹興酒 1 茶匙	1 tsp Shaoxing wine
鹽 1/2 茶匙	1/2 tsp salt
胡椒粉少許	Pinch of pepper
麻油少許	Some sesame oil

橙汁漿 | Orange sauce

橙汁 1 杯	1 cup orange juice
糖 3 湯匙	3 tbsps sugar
白醋 1 湯匙	1 tsp white vinegar
吉士粉 1 茶匙	1 tsp custard powder
橙酒 1/2 茶匙	1/2 tsp cointreau
清水 1/3 杯	1/3 cup water
鮮橙皮適量	Some orange zest

做法 | Method

1. 鴨胸解凍，洗淨抹乾，用刀在鴨肉上剞「十字」。加入醃料撈勻，待 10 分鐘，瀝乾醃汁，撲上生粉，輕輕壓實。
2. 熱鍋下油 2-3 湯匙，放入鴨胸以中細火煎至兩面微黃約八成熟，盛起瀝油。
3. 橙汁漿拌勻，以中火煮至微濃稠。
4. 放入煎鴨煮約 5 分鐘待醬汁收入鴨肉，盛起待涼，切件。

1. Thaw duck breast, rinse and pat dry, cut crosses on duck breast. Marinate duck breast for 10 minutes, drain well, dust duck breast with some corn flour, tap gently.
2. Heat a wok with 2-3 tbsps of cooking oil, pan-fry duck breast over low to medium heat until both sides are lightly brown and almost done, drain well.
3. Mix orange sauce ingredients well and cook over medium heat until slightly thickened.
4. Add duck breast into the sauce and cook for 5 minutes, set aside, slice into pieces.

入廚貼士 | Cooking Tips

- 用現成橙汁，味道和色澤比較穩定，如改用鮮榨橙汁，顏色會淺一點。
- The taste and color of the sauce are more stable by using pre-made orange juice, while the sauce color is much lighter by using fresh orange juice.

香檸煎軟鴨

Pan-fried Lemon Duck

材料 | Ingredients

急凍鴨胸 1 塊（約 300 克）
檸檬 1 個（裝飾用）
生粉適量（撲粉用）

1 pc frozen duck breast (about 300g)
1 lemon (for garnish)
Some corn flour (for dusting)

入廚貼士 | Cooking Tips

- 鴨胸很厚，為方便烹調，可預先在鴨肉上剘花和切為薄片，避免鴨肉過厚而不熟。
- As the duck breast is very thick, we can slice breast slightly or cut slits on it in advance. The duck breast can be cooked easily then.

4~6 人
Serves 4~6

30 分鐘
30 minutes

醃料 | Marinade

雞蛋 1 隻	1 egg
糖 1 茶匙	1 tsp sugar
生粉 1 茶匙	1 tsp corn flour
紹興酒 1 茶匙	1 tsp Shaoxing wine
鹽 1/2 茶匙	1/2 tsp salt
胡椒粉少許	Pinch of pepper
麻油少許	Some sesame oil

香檸汁 | Lemon sauce

檸檬汁 1 個	1 lemon (juiced)
檸檬皮 1 個	1 lemon (zest)
糖 6-8 湯匙	6-8 tbsps sugar
白醋 3 湯匙	3 tbsps white vinegar
吉士粉 1 茶匙	1 tsp custard powder
清水 1/3 杯	1/3 cup water

做法 | Method

1. 鴨胸解凍，洗淨抹乾，用刀在鴨肉上剔「十字」。加入醃料撈勻待 10 分鐘，瀝乾醃汁，撲上生粉，輕輕壓實。
2. 燒熱油，放入鴨胸以中大火炸至熟，盛起瀝油，略放涼切件。
3. 香檸汁料拌勻，以中火煮至濃稠，淋在鴨件便成。

1. Thaw duck breast, rinse and pat dry, slightly crosses on breast. Marinate for 10 minutes, drain well, dust breast with corn flour, tap gently.
2. Heat a wok with oil, deep-fry the breast over medium to high heat, drain well, cut into pieces after a while.
3. Mix lemon sauce ingredients and cook over medium heat until sauce is thickened, pour over duck breast, dish up.

4~6 人
Serves 4~6

10 分鐘
10 minutes

雜菜天婦羅
Vegetables Tempura

材料 | Ingredients

茄子 1 條	1 eggplant
番薯 1 個	1 sweet potato
泰國蘆筍數條	Several Thailand asparagus
冬菇數朵	Several dried black mushrooms
生粉適量	Some corn flour

入廚貼士 | Cooking Tips

- 炸油必須用新鮮油，才能達到效果。
- Use fresh cooking oil in order to maintain the flavor and texture.

天婦羅脆漿 | Tempura batter

天婦羅粉 1/2 杯
七味粉 1 茶匙
清水適量（按包裝指示調配）

1/2 cup tempura powder
1 tsp shichima
Some water (as per the instruction on the packing of tempura powder)

天婦羅汁 | Tempura sauce

日本醬油 1 湯匙	1 tbsp teriyaki sauce
味醂 1 湯匙	1 tbsp mirin
木魚精 1 茶匙	1 tsp Aji No Moto Hon Dashi
白蘿蔔茸 20 克	20g turnip puree
熱水 1/4 杯	1/4 cup hot water

做法 | Method

1. 蔬菜洗淨，瀝乾、切片，撲上生粉。
2. 天婦羅漿料調勻，放入蔬菜片。
3. 燒油一鍋，放入已沾天婦羅漿的蔬菜，以中火炸至酥脆金黃，盛起瀝油。
4. 天婦羅汁拌勻，與天婦羅伴吃。

1. Wash vegetables, drain well, slice, dust with corn flour.
2. Mix tempura batter, mix with vegetables.
3. Heat a wok with cooking oil, put all the vegetables in the wok, deep-fry over medium heat until golden brown and crispy, drain well.
4. Mix tempura sauce, serve with tempura vegetables.

酸甜雲吞

Deep-fried Sweet and Sour Wantons

材料 | Ingredients

雲吞皮 10-12 片	10-12 sheets wanton sheets
鯪魚肉茸 75 克	75g minced dace
三色甜椒各 1/4 個	1/4 pc each assorted color of bell peppers
菠蘿 2 片	2 slices pineapple

醃料 | Marinade

糖 1 茶匙	1 tsp sugar
生粉 1/2 茶匙	1/2 tsp corn flour
鹽 1/4 茶匙	1/4 tsp salt
胡椒粉少許	Pinch of pepper
麻油少許	Some sesame oil

4~6 人
Serves 4~6

10 分鐘
10 minutes

酸甜汁 | Sweet and sour sauce

片糖 1 塊	1 pc golden slab sugar
茄汁 1/3 杯	1/3 cup ketchup
白醋 3 湯匙	3 tbsps white vinegar
喼汁 1 湯匙	1 tbsp worcestershire sauce
生粉 1 茶匙	1 tsp corn flour
老抽 1 茶匙	1 tsp dark soy sauce
鹽 1/2 茶匙	1/2 tsp salt
清水 1/3 杯	1/3 cup water

做法 | Method

1. 鯪魚肉加入醃料拌勻待，置冰箱待 15 分鐘，放在雲吞皮上包好。
2. 燒油一鍋，放入雲吞炸至金黃，取出瀝油。
3. 三色甜椒角、洋葱、菠蘿分別切角或切粒，用少許油炒片刻盛起。
4. 熱鑊下 1 茶匙油，放入酸甜汁煮至變稠，試味。熄火，再放入甜椒、洋葱和菠蘿拌勻，盛起。

1. Mix minced dace with marinade, marinate in the freezer for 15 minutes. Wrap with wanton sheets.
2. Heat wok with cooking oil, deep-fry wantons until golden brown, drain well.
3. Cut bell peppers, onion and pineapple into wedges or cubes, saute with some cooking oil for a while, set aside.
4. Heat wok and add 1 tsp of cooking oil, add sweet and sour source, cook until suace is thickened, taste the sauce. Turn off heat, add bell peppers, onions, pineapple, toss together, serve.

入廚貼士 | Cooking Tips

- 雲吞皮很易上色變焦，必須注意油溫。
- Be alert of the cooking oil temperature as wanton sheets burnt easily.

4~6 人
Serves 4~6

10 分鐘
10 minutes

沙律蝦
Prawns with Mayonnaise

材料 | Ingredients

海中蝦 12 隻	12 medium sized prawns
沙律醬 40 克	40g mayonnaise

醃料 | Marinade

蛋白 1/2 隻	1/2 egg white
檸檬汁 1/4 個	1/4 lemon (juiced)
生粉 1 茶匙	1 tsp corn flour
油 1 茶匙	1 tsp oil
鹽 1/2 茶匙	1/2 tsp salt
糖 1/2 茶匙	1/2 tsp sugar
胡椒粉少許	Pinch of pepper

皮料 | Coating ingredients

生粉 50 克　　50g corn flour
吉士粉 10 克　　10g custard powder

做法 | Method

1. 海蝦去殼留尾，挑去蝦腸，洗淨，抹乾，並在蝦中間剘一刀，蝦尾穿過中間。
2. 把醃料拌勻，放入海蝦撈勻，醃 2-3 分鐘，取出瀝乾，沾上皮料。
3. 熱鑊下適量生油，待燒至八成滾，放入海蝦以中火作半煎炸，直至蝦約八成熟，取出瀝油。
4. 燒熱鑊後熄火，倒入沙律醬煮至微熔，立即把蝦球回鑊拌勻，便可上碟。

1. Remove prawns' shells and keep the tails, discard intestines, rinse prawns thoroughly, pat dry, cut a "Z" slit in the middle of the body, push the tail through the slit.
2. Mix marinade, marinate prawns for 2-3 minutes, drain well, coat with coating mix.
3. Heat a wok with cooking oil until 80% boil, deep-fry prawns over medium heat until nearly cooked, drain well.
4. Heat wok and then turn off heat, pour into mayonnaise, stir until it is slightly melted, toss with prawns, dish up.

入廚貼士 | Cooking Tips

- 沙律醬不要用猛火燒，以免熔掉，令蝦球很油膩。
- To avoid cooked prawns become greasy, do not cook mayonnaise over high heat, this will melt the mayonnaise.

香蒜脆蝦
Deep-fried Spicy Prawns

材料 | Ingredients

海中蝦 300 克
獨子蒜 4-5 粒
乾葱頭 2-3 個

300g medium sized prawns
4-5 cloves garlic
2-3 cloves shallot

醃料 | Marinade

生粉 1 茶匙
鹽 1/2 茶匙

1 tsp corn flour
1/2 tsp salt

4~6 人
Serves 4~6

10 分鐘
10 minutes

調味料 | Seasonings

糖 1 湯匙
老抽 1 茶匙
鹽 1/4 茶匙

1 tbsp sugar
1 tsp dark soy sauce
1/4 tsp salt

做法 | Method

1. 海中蝦去鬚、去腳，挑去蝦腸，清洗後抹乾。在蝦背上用刀剚開，加鹽撈勻，待要煎時才撲上生粉。
2. 獨子蒜去衣，切薄片。乾葱頭去衣，切薄片。
3. 熱鑊下適量油，待燒至八成滾，放入獨子蒜片和乾葱頭片，炸至金黃，盛起。再放入海中蝦炸熟兼呈微金黃，盛起。
4. 原鑊留 1 湯匙油，把海蝦回鑊，倒入調味炒勻，熄火，放入炸好的蒜片和乾葱片拌勻，上碟。

1. Remove prawns' antennae and legs, discard intestines, wash prawns thoroughly, pat dry. Cut a "Z" slit on the back of the body, season with salt, coated with corn flour before pan-frying.
2. Peel garlic, cut into thin pieces. Peel shallot, cut into thin slices.
3. Heat a wok with oil until 80% boil, deep-fry garlic and shallot until golden brown, set aside. Deep-fry prawns until slightly golden brown, set aside.
4. Heat the same wok with 1 tbsp of cooking oil, add prawns, add seasonings, turn off heat and toss with garlic and shallot, dish up.

4~6 人
Serves 4~6

40 分鐘
40 minutes

酥炸蝦丸
Deep-fried Prawn Balls

材料 | Ingredients

蝦仁（硬殼蝦）300 克
墨魚膠 75 克

300g prawns (hard-shelled)
75g cuttlefish paste

醃料 | Marinade

蛋白 1 隻	1 egg white
生粉 1 湯匙	1 tbsp corn flour
鹽 1 茶匙	1 tsp salt
糖 1 茶匙	1 tsp sugar
油 1 茶匙	1 tsp oil
胡椒粉適量	Pinch of pepper

蘸汁 | Dipping sauce

檸檬角數件
沙律醬適量

Some lemon wedges
Some mayonnaise

做法 | Method

1. 蝦仁用鹽水清洗，加生粉撈勻。然後用清水沖淨，再用廚紙吸乾水份。放入保鮮袋，用刀大力拍扁。
2. 把墨魚膠、蝦仁茸和醃料撈勻，順時針方向大力攪至起膠有彈力。放冰箱冷藏 30 分鐘。
3. 炸油燒熱至八成滾，用手唧蝦丸。放入油中以中火炸至金黃，取出瀝油。伴以沙律醬和檸檬角享用。

1. Rinse prawns with brine, mix with corn flour. Rinse prawns again, pat dry with kitchen papers. Put in a plastic bag, smash prawns with the blade of a knife.
2. Marinate mixed cuttlefish paste and prawn paste, stir vigorously in clockwise direction until the mixture becomes elastic. Place in a freezer for 30 minutes.
3. Heat cooking oil in a wok until 80% boil, hand squeezes prawn balls. Deep-fry prawn balls until golden brown, drain well. Serve with mayonnaise and lemon wedges.

入廚貼士 | Cooking Tips

- 蝦仁要徹底清洗，把表面的黏液去掉，再吸乾水份，蝦膠才能清爽兼有彈力。
- To ensure prawn paste is rinse, crunchy and chewy, we need to rinse prawns thoroughly, remove fluid on the surface and pat dry.

炸蟹箝

Deep-fried Crab Claws

材料 | Ingredients

蟹箝 6 隻	6 crab claws
蝦仁（硬殼蝦）300 克	300g prawns (hard shelled prawns)
墨魚膠 75 克	75g cuttlefish paste

醃料 | Marinade

蛋白 1 隻	1 egg white
生粉 1 湯匙	1 tbsp corn flour
鹽 1 茶匙	1 tsp salt
糖 1 茶匙	1 tsp sugar
油 1 茶匙	1 tsp oil
胡椒粉適量	Pinch of pepper

入廚貼士 | Cooking Tips

- 蟹箝蒸熟後立即浸入冰水，待片刻，用裹上毛巾的菜刀輕拍裂，便可輕易拆殼。
- There is an easier way to remove the shells of crab claws. After putting steamed crab crabs in ice water right after steaming process for a while, crack the crab claws with a knife which is wrapped by a towel.

4~6 人
Serves 4~6

10 分鐘
10 minutes

吉列料 | Cutlets ingredients

雞蛋 1 隻（打散）	1 egg (whisked)
粗麵包糠 1 杯	1 cup coarse breadcrumbs
麵粉 1/3 杯	1/3 cup flour

蘸汁 | Dipping sauce

檸檬角數件	Some lemon wedges
喼汁適量	Some worcestershire sauce

做法 | Method

1. 蝦仁用鹽水洗淨，加生粉撈勻，用清水沖洗，然後用廚紙吸乾水份。放入保鮮袋，用刀大力拍扁。加入墨魚膠和醃料撈勻，順時針方向大力攪至起膠，放冰箱冷藏 30 分鐘。
2. 蟹箝洗淨，以大火蒸 10 分鐘。拆殼，保留最尖端部份，不要弄破，再裹上蝦膠。
3. 分別順序沾上麵粉、雞蛋液和麵包糠，輕輕按實。
4. 熱鑊下適量油，待燒至八成滾，放入蟹箝中以中火炸至金黃，取出瀝油。伴以沙律醬或喼汁，配檸檬角享用。

1. Wash prawns with brine, mix with corn flour, wash well, pat dry with kitchen papers. Put in a zipper bag, smash vigorously with the blade of a knife. Mix with cuttlefish paste and marinade, stir vigorously in the clockwise direction until mixture becomes elastic, place in the freezer for 30 minutes.
2. Wash crab claws and steam over high heat for 10 minutes. Remove shells, keep the tip of shells in good shape, do not break them, stuff and coat shells with prawn paste.
3. Coat stuffed shells with flour, egg and breadcrumbs in order, tag gently.
4. Heat a wok with oil until 80% boil, deep-fry crab claws over medium heat until golden brown, drain well. Serve with mayonnaise or worcestershire sauce and lemon wedges.

4~6 人
Serves 4~6

20 分鐘
20 minutes

粟米魚塊

Fish Fillet with Sweet Corn Sauce

材料 | Ingredients

魚柳 1 條（約 450 克）	1 fish fillet (about 450g)
粟米羹 1/2 罐（約 200 克）	1/2 can sweet corn cream soup (about 200g)
雞蛋 1 隻（打散）	1 egg (whisked)
清水 1/4 杯	1/4 cup water

醃料 | Marinade

檸檬汁 1 個	1 lemon (juiced)
雞蛋 1 隻	1 egg
生粉 2 茶匙	2 tsps corn flour
油 2 茶匙	2 tsps oil
鹽 1 茶匙	1 tsp salt
糖 1 茶匙	1 tsp sugar
胡椒粉少許	Pinch of pepper

調味料 | Seasonings

鹽 1/2 茶匙
糖 1/2 茶匙
胡椒粉少許

1/2 tsp salt
1/2 tsp sugar
Pinch of pepper

皮料 | Coating ingredients

雞蛋 1 隻（打匀）
生粉 1 杯

1 egg (whisked)
1 cup corn flour

做法 | Method

1. 魚柳解凍洗淨，抹乾切塊。放入醃料撈匀，醃約 10 分鐘，取出瀝乾。然後沾上雞蛋液和生粉，輕輕按實。
2. 炸油燒熱至八成滾，放入魚塊，以中火炸至金黃色，取出瀝油。再放在廚紙上吸油，上碟。
3. 罐裝粟米羹倒入小鍋煮滾，加入調味，熄火。拌入雞蛋液，淋在魚塊上。

1. Thaw fish fillet, rinse, pat dry and cut into pieces. Marinate them for 10 minutes, drain well. Coat with whisked egg and corn flour, tap gently.
2. Heat cooking oil until 80% boil, deep-fry fillet over medium heat until golden brown, drain well. Absorb excess oil with kitchen paper, set aside.
3. Boil sweet corn cream soup in a small saucepan, season, turn off heat. Mix with whisked egg, stir, pour over fish fillet. Serve.

入廚貼士 | Cooking Tips

- 粟米羹帶黏度，而雞蛋液因加熱會凝結，所以不用勾芡汁。蛋絲要滑嫩，必須熄火才加進羹內，否則蛋絲會粗糙。
- The sweet corn soup is creamy and whisked egg will be coagulated when heated, therefore, it is not necessary to add thickening sauce. To ensure egg shreds are smooth and tender in texture, add whisked egg into sauce when heat has turned off.

炸魚手指

Deep-fried Fish Finger

材料 | Ingredients

青衣魚 / 龍脷柳 1 塊（約 450 克）	1 pc parrot fish/sole fillet (about 450g)
番茄 1-2 個（切片）	1-2 tomatoes (sliced)
炸油 2 杯	2 cups cooking oil

醃料 | Marinade

檸檬汁 1 個	1 lemon (juiced)
雞蛋 1 隻	1 egg
生粉 2 茶匙	2 tsps corn flour
油 2 茶匙	2 tsps oil
鹽 1 茶匙	1 tsp salt
糖 1 茶匙	1 tsp sugar
胡椒粉少許	Pinch of pepper

4~6 人
Serves 4~6

15 分鐘
15 minutes

吉列料 | Cutlets ingredients

雞蛋 1 隻（打散）
粗麵包糠 1 杯
麵粉 1/3 杯

1 egg (whisked)
1 cup coarse breadcrumbs
1/3 cup flour

蘸汁 | Dipping sauce

茄汁適量
喼汁適量

Some ketchup
Some worcestershire sauce

做法 | Method

1. 龍脷魚解凍洗淨，用布抹乾水份，切成 2.5 厘米寬 ×10 厘米長的魚條。
2. 用醃料撈勻魚條，醃約 10 分鐘，取出瀝乾。
3. 順序把魚條沾上麵粉、雞蛋液和麵包糠，輕輕按實。
4. 燒油 2 杯，熱至八成滾，放入魚條，以中火炸至變微黃色。
5. 改猛火炸至金黃，取出，瀝油，再放在廚紙上吸油。可伴蘸汁和番茄片享用。

1. Thaw sole, rinse and pat dry with a towel, cut it into 2.5 cm wide x 10 cm long strips.
2. Marinate fish strips for 10 minutes, remove and drain well.
3. Dip fish strips with flour, whisked egg and breadcrumbs in order, tap gently.
4. Heat a wok with 2 cups of oil until 80% boil, deep-fry fish strips over medium heat until slightly golden brown.
5. Turn to high heat and deep-fry fish strips till golden brown. Remove, drain well and absorb excess oil by kitchen paper. Serve with dipping sauce and sliced tomatoes.

入廚貼士 | Cooking Tips

- 魚柳要揀肉厚而平均的，還要每塊盡量大小相若。魚柳沾上吉列料後必須按實，炸時才不易掉落碎屑在炸油內，弄污炸油。
- Choose fillets which are thick and with equal size. Make sure to tag the strips after coating with cutlet ingredients. This will prevent breadcrumbs mixture fall apart and spoil the cooking oil.

4~6 人
Serves 4~6

10 分鐘
10 minutes

香茅炸豬扒

Deep-fried Pork Chops with Lemongrass

材料 | Ingredients

豬扒 2 塊
鮮香茅 1-2 支（切碎）
乾葱頭 2-3 粒（切片）
生粉適量（撲面用）

2 pcs pork chops
1-2 stalks fresh lemongrass (chopped)
2-3 cloves shallot (slices)
Some corn flour (for dusitng)

入廚貼士 | Cooking Tips

- 醃料有魚露調味，就不用下鹽了，否則會太鹹。
- As fish sauce is salty, it is not necessary to season with salt.

醃料 | Marinade

雞蛋 1 隻	1 egg
鮮香茅碎 1 湯匙	1 tbsp fresh chopped lemongrass
魚露 1 茶匙	1 tsp fish sauce
糖 1 茶匙	1 tsp sugar
生粉 1 茶匙	1 tsp corn flour
油 1 茶匙	1 tsp oil
紹興酒 1/2 茶匙	1/2 tsp Shaoxing wine

芡汁 | Thickening

糖 2 茶匙	2 tsps sugar
檸檬汁 1 茶匙	1 tsp lemon juice
生粉 1 茶匙	1 tsp corn flour
老抽 1/2 茶匙	1/2 tsp dark light sauce
魚露 1/2 茶匙	1/2 tsp fish sauce
清水 4 湯匙	4 tbsps water

做法 | Method

1. 豬扒解凍，洗淨抹乾，用刀背稍拍。加入醃料撈勻待 15 分鐘，撲上生粉輕輕按實。
2. 燒熱炸油，把豬扒以中火炸至八成熟，轉大火上色，盛起瀝油。
3. 熱鑊下油，放入芡汁煮至濃稠，加進香茅碎和乾葱片炒勻。

1. Thaw pork chops, rinse and pat dry. Pound pork chops with the back of a knife. Marinate for 15 minutes, dust with corn flour and tap gently.
2. Heat a wok with cooking oil, deep-fry pork chops over medium heat until 80% cooked. Turn to high heat to brown pork chops, drain well.
3. Add cooking oil in a preheated wok, cook thickening sauce until thickened, stir-fry pork chops with lemongrass and shallot.

吉列豬扒

Pork Chop Cutlet

材料 | Ingredients

豬扒 2-3 塊
2-3 pcs pork chops

入廚貼士 | Cooking Tips

- 吉列菜餚會比較油膩，需要用廚紙吸油，保持乾爽酥脆。
- Cutlet dishes are greasy, it is better to absorb excess grease by kitchen paper in order to preserve crunchy texture.

4~6 人
Serves 4~6

15 分鐘
15 minutes

醃料 | Marinade

糖 1 茶匙	1 tsp sugar
生粉 1 茶匙	1 tsp corn flour
油 1 茶匙	1 tsp oil
鹽 1/2 茶匙	1/2 tsp salt
紹興酒 1/2 茶匙	1/2 tsp Shaoxing wine

吉列料 | Cutlet ingredients

雞蛋 1 隻（打勻）	1 egg (whisked)
粗麵包糠 1 杯	1 cup coarse breadcrumbs
麵粉 1/3 杯	1/3 cup flour

伴食 | Condiments

檸檬數角	Some lemon wedges
沙律醬適量	Some mayonnaise
日式燒汁適量	Some teriyaki sauce

做法 | Method

1. 豬扒解凍，洗淨抹乾。用刀背稍拍，加入醃料撈勻待 15 分鐘。
2. 順序沾上麵粉、雞蛋液和麵包糠，輕輕按實。
3. 燒熱炸油，放入豬扒以中火炸至八成熟。轉大火上色，盛起瀝油。
4. 待片刻，斬成數件，伴以檸檬角上桌。

1. Thaw pork chops, rinse and pat dry. Pound pork chops with the back of a knife, marinate for 15 minutes.
2. Coat pork chops with flour, whisked egg and breadcrumbs in order, tap gently.
3. Heat a wok with cooking oil, deep-fry pork chops over medium heat until 80% cooked. Turn to high heat to brown pork chops, drain well.
4. Cut pork chops into pieces, serve with lemon wedges.

4~6 人
Serves 4~6

10 分鐘
10 minutes

自家燒肉
Homemade Roast Pork

材料 | Ingredients

連皮豬腩肉 1 條（約 600 克）

1 pc roast pork belly with skin (about 600g)

入廚貼士 | Cooking Tips

- 豬腩肉刺孔，較容易做到豬皮爆脆效果。
- By pricking holes on the pork belly can help to create crispy pork crackling.

醃料 | Marinade

醬油 2 湯匙
糖 2 湯匙
燒烤醬 2 湯匙
生粉 1 湯匙
紅酒 1 湯匙

2 tbsps soy sauce
2 tbsps sugar
2 tbsps BBQ sauce
1 tbsp corn flour
1 tbsp red wine

做法 | Method

1. 先把豬腩肉的豬皮用粗針刺孔，再把所有醃料及豬腩肉撈勻醃至 5 小時以上。
2. 用錫紙把豬肉包好，隔水蒸熟，瀝乾水份。
3. 熱鑊燒油，豬皮向下先炸脆。翻轉，再炸豬腩肉 3-5 分鐘。
4. 用工具吊起豬腩肉，淋上滾油，直至熟透。

1. Prick holes on the skin of pork belly with a thick pin, marinate pork belly for at least 5 hours.
2. Wrap pork belly with aluminum foil, steam until done, drain well.
3. Heat a wok with cooking oil, deep-fry pork belly with the skin face down till the skin becomes crispy. Turn the meat over, deep-fry again for 3-5 minutes.
4. Hang pork belly, pour hot oil over it, repeat until cooked.

福州炸五香肉卷

Fuzhou Deep-fried Spicy Spring Rolls

材料 | Ingredients

半肥瘦豬肉 150 克	150 g pork (with half lean meat and half fatty meat)
馬蹄肉 60 克（剁碎）	60g water chestnuts (chopped)
葱 2 條（切粒）	2 stalks spring onion (chopped)
腐皮 1 塊	1 beancurd sheet
薯粉 40 克	40g potato flour

醃料 | Marinade

雞蛋 1 隻	1 egg
鹽 2 茶匙	2 tsps salt
清水 1 茶匙	1 tsp water

4~6 人
Serves 4~6

10 分鐘
10 minutes

調味料｜Seasonings

糖 2 茶匙	2 tsps sugar
生抽 1 茶匙	1 tsp light soy sauce
五香粉 1/2 茶匙	1/2 tsp five spicy powder
胡椒粉適量	Pinch of pepper

麵粉漿｜Batter

麵粉 2 湯匙	2 tbsps flour
清水 3 湯匙	3 tbsps water

做法｜Method

1. 半肥瘦豬肉切粗粒，加入醃肉料大力攪匀。然後加入調味料和薯粉拌匀，再加入馬蹄粒。拌匀後置雪櫃中冷凍 2 小時。
2. 麵粉漿料調匀成濃糊狀。
3. 腐皮切成四等份，分別放入肉餡，捲成長卷形。邊緣塗上粉漿，按緊收口。
4. 將肉卷置在蒸籠上，隔水以大火蒸 3 分鐘至半熟，取出。
5. 燒油一鍋，以中慢火炸至金黃色。隔去油分，用吸油紙吸油，切件上碟。

1. Cut pork into coarse dice, add marinade and stir well vigorously. Add seasoning and potato flour and mix well, then add water chestnut dice and mix well. Put into refrigerator for 2 hours.
2. Mix batter well.
3. Cut beancurd sheet into 4 equal portions, add fillings and roll up into spring roll shape. Brush the ends with batter and press the ends firmly.
4. Put spring rolls onto a steamer and steam over high heat for 3 minutes until half-cooked, dish up.
5. Heat a wok of oil and deep-fry spring rolls over medium heat until golden. Drain excess oil and absorb by kitchen paper. Cut into pieces and dish up.

4~6 人
Serves 4~6

10 分鐘
10 minutes

肉碎脆米粉
Deep-fried Rice Noodles with Minced Pork

材料 | Ingredients

粉絲 / 米粉 40 克	40g vermicelli/ rice noodles
瘦豬肉 150 克	150g lean pork meat
葱粒 1 湯匙	1 tbsp chopped spring onion
蒜茸 1 茶匙	1 tsp minced garlic
辣椒碎 1 茶匙	1 tsp chopped chili

醃料 | Marinade

糖 1 茶匙	1 tsp sugar
生粉 1 茶匙	1 tsp corn flour
油 1 茶匙	1 tsp oil
鹽 1/2 茶匙	1/2 tsp salt
紹興酒 1/2 茶匙	1/2 tsp Shaoxing wine

芡汁 | Thickening

生粉 1 茶匙	1 tsp corn flour
糖 1/2 茶匙	1/2 tsp sugar
鹽 1/4 茶匙	1/4 tsp salt
清水 3-4 湯匙	3-4 tbsps water
胡椒粉少許	Pinch of pepper
麻油少許	Some sesame oil

做法 | Method

1. 瘦豬肉洗淨，切粒，加入醃料撈勻，待 10 分鐘。
2. 燒滾油一鍋，至開始冒煙，放入粉絲或米粉炸至脹大。翻轉，保持色澤潔白，而每條粉絲皆充分膨脹，盛起。
3. 熱鑊下 1-2 茶匙油，放入蒜茸和辣椒碎爆香。倒進瘦豬肉粒快炒至熟並微焦。加入芡汁煮稠，淋在炸粉絲上，撒上葱粒便成。

1. Rinse pork, cut into cubes, marinate for 10 minutes.
2. Heat a wok with cooking oil over high heat, when oil starts to burn, add vermicelli or rice noodles and let them swell. Turn vermicelli over, make sure all vermicelli swell, set aside.
3. Heat wok with 1-2 tsps of cooking oil, stir-fry garlic and chili. Add lean meat and stir-fry quickly, do not burn the pork. Add thickening sauce and cook till thickened, pour over vermicelli, garnish with chopped spring onion.

入廚貼士 | Cooking Tips

- 粉絲不要洗。投入少量粉絲於炸油中，能立即冒起，表示油溫適合，太快或太慢升起都不適合。
- Do not wash vermicelli. To test the temperature of cooking oil, drop a small pinch of vermicelli into the oil. If vermicelli rise immediately, oil temperature is suitable. The oil temperature is not suitable if vermicelli rise too slow or too fast.

生炒排骨
Sweet and Sour Pork

材料 | Ingredients

排骨 450 克	450g pork ribs
三色椒各 1/2 個	1/2 pc each assorted color of bell peppers
洋葱 1/2 個	1/2 onion
菠蘿 2 片	2 slices pineapple
生粉適量	Some corn flour

醃料 | Marinade

糖 1 茶匙	1 tsp sugar
生粉 1 茶匙	1 tsp corn flour
油 1 茶匙	1 tsp oil
鹽 1/2 茶匙	1/2 tsp salt
紹興酒 1/2 茶匙	1/2 tsp Shaoxing wine
胡椒粉少許	Pinch of pepper
麻油少許	Some sesame oil
雞蛋 1 隻（後下）	1 egg (use later)

酸甜汁 | Sweet and sour sauce

片糖 1 塊	1 pc golden slab sugar
茄汁 1/3 杯	1/3 cup ketchup
白醋 3 湯匙	3 tbsps white vinegar
喼汁 1 湯匙	1 tbsp worcestershire sauce
生粉 1 茶匙	1 tsp corn flour
老抽 1 茶匙	1 tsp dark soy sauce
鹽 1/2 茶匙	1/2 tsp salt
清水 1/3 杯	1/3 cup water

做法 | Method

1. 排骨洗淨抹乾，加入醃料拌勻待 30 分鐘。拌入雞蛋，撲上生粉輕輕捏實。
2. 燒熱炸油，把排骨炸至八成熟盛起，撈去油下污物。燒熱後再把排骨回鑊，炸至金黃，取出瀝油。
3. 三色椒、洋葱、菠蘿分別切角或切粒，用少許油炒片刻盛起。
4. 熱鑊下 1 茶匙油，放入酸甜汁煮至變稠，試味，熄火。再放入三色椒、洋葱和菠蘿拌勻，盛起。

1. Rinse pork ribs and pat dry, marinate for 30 minutes. Coat with whisked egg, dust with corn flour and tap gently.
2. Heat a wok with cooking oil, deep-fry the pork ribs until 80% cooked, set aside, remove debris in the cooking oil. Reheat wok and deep-fry pork ribs again until golden brown, drain well.
3. Cut bell peppers, onion and pineapple into wedges or cubes, stir-fry with some cooking oil, set aside.
4. Heat wok with 1 tsp of cooking oil, cook sweet and sour sauce until thickened, taste, turn off heat. Toss bell peppers, onion and pineapple with sauce, dish up.

入廚貼士 | Cooking Tips

- 要享用酥脆的排骨，就不要放進酸甜汁燴煮，分開上桌。
- To preserve the crunchy texture of pork ribs, serve the sauce separately instead of cooking together.

4~6 人
Serves 4~6

10 分鐘
10 minutes

酥炸雞球

Crispy Spicy Deep-fried Chicken Wings

材料 | Ingredients

雞翼 10-12 隻
10-12 chicken wings

醃料 | Marinade

雞蛋 1 隻	1 egg
糖 1 茶匙	1 tsp sugar
生粉 1 茶匙	1 tsp corn flour
油 1 茶匙	1 tsp oil
鹽 1/2 茶匙	1/2 tsp salt
紹興酒 1/2 茶匙	1/2 tsp Shaoxing wine
胡椒粉少許	Pinch of pepper
麻油少許	Some sesame oil

皮料 | Coating ingredients

麵粉 3 湯匙	3 tbsps flour
吉士粉 1 湯匙	1 tbsp custard powder
生粉 1 湯匙	1 tbsp corn flour

伴食 | Condiments

檸檬數角	Some lemon wedges
泰式雞醬適量	Some Thai chicken sauce

做法 | Method

1. 雞翼解凍，用鹽擦後清洗，抹乾，起骨。加入醃料撈勻待 10 分鐘，瀝乾。
2. 皮料拌勻，塗在雞翼上，輕輕按實。
3. 熱鑊燒一鍋油至八成滾，放進雞翼以中火炸熟。轉大火炸至呈金黃色，撈出瀝油。可配泰式雞醬和檸檬角享用。

1. Thaw chicken wings, rub chicken wings with salt, wash and pat dry, debone. Marinate for 10 minutes, drain well.
2. Mix coating ingredients, coat chicken wings, tap gently.
3. Heat a wok with cooking oil until 80% boil, deep-fry chicken over medium heat until cooked. Turn to high heat to deep-fry chicken wings until golden brown, drain well. Serve with Thai chicken sauce and lemon wedges.

入廚貼士 | Cooking Tips

- 雞翼起骨，可去掉幼骨，獨留粗骨，然後把雞肉翻起，效果也不錯。
- There is another way to prepare chicken wings which can also give a nice presentation. Keep the thicker chicken bone and discard the thinner one during the deboning process, and then turn chicken meat over.

椒鹽蒜香炸雞翼

Deep-fried Crispy Chicken Wings with Chili Salt

材料 | Ingredients

雞翼 10-12 隻	10-12 chicken wings
辣椒 1 隻（切碎）	1 chili (chopped)
蒜茸 2 茶匙	2 tsps minced garlic
生粉適量（撲面用）	Some corn flour (for dusting)

入廚貼士 | Cooking Tips

- 蒜茸常常用，可預先剁碎，再淋上熟油，放在冰箱隨時取用。
- As minced garlic is used frequently, we can chop some garlic and cover with cooked oil and store in the fridge for later use.

4~6 人
Serves 4~6

10 分鐘
10 minutes

醃料 | Marinade

蛋白 1 隻	1 egg white
糖 1 茶匙	1 tsp sugar
生粉 1 茶匙	1 tsp corn flour
油 1 茶匙	1 tsp oil
鹽 1/2 茶匙	1/2 tsp salt
紹興酒 1/2 茶匙	1/2 tsp Shaoxing wine
胡椒粉少許	Pinch of pepper
麻油少許	Some sesame oil

調味料 | Seasonings

蒜鹽 1 茶匙	1 tsp garlic salt
七味粉 1/2 茶匙	1/2 tsp shichimi

做法 | Method

1. 雞翼解凍，用鹽擦後清洗，抹乾。加入醃料撈勻待 10 分鐘，瀝乾，撲上生粉。
2. 熱鑊下適量生油，待燒至八成滾，放進雞翼以中火炸熟。轉大火上色，撈出瀝油。
3. 熱鑊爆香辣椒碎和蒜茸，倒入雞翼和調味炒勻，上碟。

1. Thaw chicken wings, rub with salt, wash and pat dry. Marinate for 10 minutes, drain well, dust with corn flour.
2. Heat a wok with cooking oil until 80% boil, deep-fry chicken wings over medium heat until cooked. Turn to high heat to brown chicken wings, drain well.
3. Heat wok, sauté chili and minced garlic, toss with chicken wings and seasonings, dish up.

4~6 人
Serves 4~6

10 分鐘
10 minutes

香草牛油炸雞翼

Deep-fried Herbal Chicken Wings

材料 | Ingredients

雞翼 10-12 隻
生粉適量（撲面用）

10-12 pcs chicken wings
Some corn flour (for dusting)

入廚貼士 | Cooking Tips

- 雞翼解凍後可用鹽或檸檬汁擦洗，去除雪藏味。
- To remove peculiar smell of frozen chicken wings, rub defrosted chicken wings with salt or lemon juice first and then wash thoroughly.

醃料 | Marinade

蛋白 1 隻
糖 1 茶匙
生粉 1 茶匙
油 1 茶匙
鹽 1/2 茶匙
紹興酒 1/2 茶匙
胡椒粉少許
麻油少許

1 egg white
1 tsp sugar
1 tsp corn flour
1 tsp oil
1/2 tsp salt
1/2 tsp Shaoxing wine
Pinch of pepper
Some sesame oil

香草牛油 | Herbal butter

無鹽牛油 1 湯匙（放軟）
雜香草 1 茶匙
蒜茸 1 茶匙
蒜鹽 1 茶匙

1 tbsp unsalted butter (soften)
1 tsp mixed herbs
1 tsp minced garlic
1 tsp salt garlic

做法 | Method

1. 雞翼解凍，用鹽擦後清洗，抹乾。加入醃料撈勻待 10 分鐘，瀝乾，撲上生粉。
2. 熱鑊下適量生油，待燒至八成滾，放進雞翼以中火炸熟。轉大火上色，撈出瀝油。
3. 香草牛油料拌勻，放入已預熱的鑊內，倒進雞翼炒勻。

1. Thaw chicken wings, rub with salt, wash and pat dry. Marinate for 10 minutes, drain well, dust with corn flour.
2. Heat a wok with cooking oil until 80% boil, deep-fry chicken wings over medium heat until cooked. Turn to high heat to brown chicken wings, drain well.
3. Mix herbal butter ingredients, put mixture in a pre-heated wok, sauté chicken wings.

酥炸苔條鴨胸

Deep-fried Crispy Duck Breast

材料 | Ingredients

煙鴨胸 1 塊（約 300 克）
生粉適量（撲粉用）

1 pc smoked duck breast (about 300g)
Some corn flour (for dusting)

醃料 | Marinade

糖 1 茶匙
檸檬汁 1 茶匙

1 tsp sugar
1 tsp lemon juice

4~6 人
Serves 4~6

30 分鐘
30 minutes

脆漿 | Batter

脆炸粉 1 杯
海苔條 1 小撮（略浸）
清水適量（按產品商建議）

1 cup crispy deep-frying powder
Several seaweed strips (slightly soaked)
Some water (per instruction on the packing)

蘸汁 | Dipping sauce

喼汁適量
沙律醬適量

Some worcestershire sauce
Some mayonnaise

做法 | Method

1. 煙鴨胸用刀輕剔數下，加入醃料抹勻，撲點生粉，備用。
2. 脆炸粉與清水調勻，放入海苔條拌勻。
3. 燒油一鍋至八成滾，放入已沾脆漿的鴨胸，油炸至酥脆和呈金黃。
4. 撈出鴨胸瀝油，待涼切件，可伴喼汁和沙律醬享用。

1. Lightly slit duck breast with a few crosses, marinate, dust with some corn flour, set aside.
2. Mix crispy deep-frying powder with water, then mix seaweed strips.
3. Heat a wok with cooking oil until 80% boil, coat duck breast with crispy deep-frying mixture, deep-fry duck breast until crispy and golden brown.
4. Drain duck breast, slice into pieces after a while, serve with worcestershire sauce and mayonnaise.

入廚貼士 | Cooking Tips

- 煙鴨胸的味道濃和鹹，用點檸檬汁和砂糖可中和鴨肉的味道。
- Since duck breast is salty and has strong flavor, add some lemon juice and sugar to improve the taste.

編著
何美好

編輯
Pheona Tse　Kitty Chan

美術設計
YU Cheung

排版
辛紅梅

翻譯
Jennie

攝影
Lasso Adv. Agency

出版者
萬里機構出版有限公司
香港鰂魚涌英皇道1065號東達中心1305室
電話：2564 7511
傳真：2565 5539
電郵：info@wanlibk.com
網址：http://www.wanlibk.com
http://www.facebook.com/wanlibk

萬里機構

萬里 Facebook

發行者
香港聯合書刊物流有限公司
香港新界大埔汀麗路36號
中華商務印刷大廈3字樓
電話：2150 2100
傳真：2407 3062
電郵：info@suplogistics.com.hk

承印者
美雅印刷製本有限公司

出版日期
二零一八年九月第一次印刷

ISBN 978-962-14-6743-0